Géza Tóth
Áron Kincses
Zoltán Nagy

European Spatial Structure

Géza Tóth
Áron Kincses
Zoltán Nagy

European Spatial Structure

LAP LAMBERT Academic Publishing

Impressum / Imprint

Bibliografische Information der Deutschen Nationalbibliothek: Die Deutsche Nationalbibliothek verzeichnet diese Publikation in der Deutschen Nationalbibliografie; detaillierte bibliografische Daten sind im Internet über http://dnb.d-nb.de abrufbar.
Alle in diesem Buch genannten Marken und Produktnamen unterliegen warenzeichen-, marken- oder patentrechtlichem Schutz bzw. sind Warenzeichen oder eingetragene Warenzeichen der jeweiligen Inhaber. Die Wiedergabe von Marken, Produktnamen, Gebrauchsnamen, Handelsnamen, Warenbezeichnungen u.s.w. in diesem Werk berechtigt auch ohne besondere Kennzeichnung nicht zu der Annahme, dass solche Namen im Sinne der Warenzeichen- und Markenschutzgesetzgebung als frei zu betrachten wären und daher von jedermann benutzt werden dürften.

Bibliographic information published by the Deutsche Nationalbibliothek: The Deutsche Nationalbibliothek lists this publication in the Deutsche Nationalbibliografie; detailed bibliographic data are available in the Internet at http://dnb.d-nb.de.
Any brand names and product names mentioned in this book are subject to trademark, brand or patent protection and are trademarks or registered trademarks of their respective holders. The use of brand names, product names, common names, trade names, product descriptions etc. even without a particular marking in this work is in no way to be construed to mean that such names may be regarded as unrestricted in respect of trademark and brand protection legislation and could thus be used by anyone.

Coverbild / Cover image: www.ingimage.com

Verlag / Publisher:
LAP LAMBERT Academic Publishing
ist ein Imprint der / is a trademark of
OmniScriptum GmbH & Co. KG
Heinrich-Böcking-Str. 6-8, 66121 Saarbrücken, Deutschland / Germany
Email: info@lap-publishing.com

Herstellung: siehe letzte Seite /
Printed at: see last page
ISBN: 978-3-659-64559-4

Content

Introduction

Our study aims at describing the spatial structure of Europe with spatial moving average, potential model and the bidimensional regression analysis based on gravity model. Many theoretical and practical works aim at describing the spatial structure of Europe. Partly zones, axes and formations, partly polycentric models appear in the literature. We illustrate their variegation by listing, without any claim to completeness (since that could be the subject of another study), a part of them. Based on our examinations, the engraving of the structures that we described can be seen. The position of the core area of EU countries clearly justifies the banana shape and in relation to it, the catching up regions take shape in several areas.

Spatial structure of Europe

There have been many attempts to reveal and visualise the varied economic and social structural image of Europe in the last decades. These models attempt to demonstrate the determinant elements of the geographic space, the complex systems among them and the characteristics of this space structure. Spatial structural visualizations are differentiated along two approaches: one including zones, axes and formations and the other one including polycentric models.

The first provocative form was published in the study of Brunet (1989) as the "European Backbone". Later it was called by its popular name "Blue Banana". The authors drew a banana-shaped form to visualise the economic core area approximately from Liverpool to Nice (or from London to Milan). (Figure 1). Our figures present – without any claim to completeness – the approaches that we consider to be the most important ones.

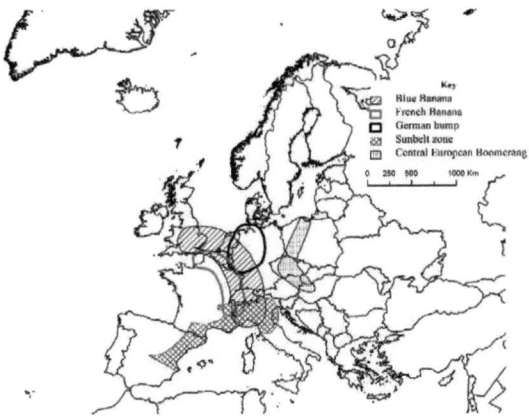

Figure 1: Spatial structure models I (source: own compilation based on Brunet 1989, Gorzelak 2012; Kunzmann 1992; Schatzl 1993; Hospers 2002).

A form similar to the banana can also be found in East-Central Europe called the "Central European Boomerang" (see Figure 1). According to Gorzelak (2012), the determinant areas of this form – stretching from Gdansk to Budapest and including Poznan, Wroclaw, Prague and the triangle of Vienna-Bratislava-Budapest – are the capitals, the real places of development.

Figure 2: Spatial structure models II (source: own
compilation based on van der Meer 1998 and ESDP 1999).

Further forms have appeared in the literature, such as the
"Red Octopus", the body and the Western arms of which
stretch between Birmingham and Barcelona toward Rome
and Paris. It stretches toward Copenhagen-Stockholm
(Helsinki) to the North and toward Berlin-Poznan-Warsaw
and Prague-Vienna-Budapest to the East (van der Meer 1998)
(Figure 2). Unlike earlier visualizations, this form includes
the group of developed zones and their core cities,
highlighting the possibilities to decrease spatial differences
in this way as well by visualizing polycentricity and
"eurocorridors" (Szabó 2009). The "Blue Star" is a bit similar
to this form. In spite of the fact that it has not become as
popular, the "Blue Star" also indicates the directions of
development and the dynamic areas with the visualization of
arrows and therefore makes future references possible
(Dommergues 1992) (Figure 3).

Figure 3: Spatial structure models III (source: own
compilation based on Dommergues 1992).

The "European Pentagon" (Figure 2) is the region defined by
London-Paris-Milan-Munich-Hamburg in the European
Spatial Development Perspective (ESDP) in 1999.

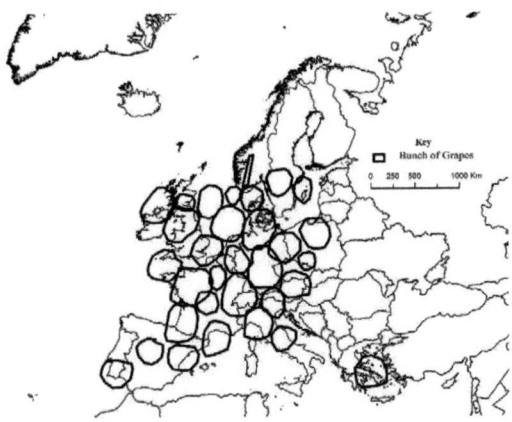

Figure 4: Spatial structure models IV (source: own
compilation based on Kunzmann 1992).

The other important group in the visualization of spatial structure highlights urban development, the dynamic change of urban areas and the polycentric spatial structure (one of them can be seen on Figure 4) (Szabó 2009). Kunzmann and Wegener (Kunzmann and Wegener 1991; Kunzmann 1992, 1996; Wegener and Kunzmann 1996) did not agree with the spatial description of the "Blue Banana" and other forms. They believe that the polycentric structure of our continent is determined by the metropolitan regions (which are situated not only within the "Blue Banana"), situated in a "Bunch of Grapes" shape. After this, polycentricity became an increasingly popular idea and one of the key elements of ESDP 1999. It also has an increasingly important role in the European cohesion policy (Faludi 2005; Kilper 2009). At the same time, however, critical statements appear against this kind of approach of planning, for example from the point of view of economic efficiency or sustainable development (Vandermotten et al. 2008).

This structure is reflected in the so-called MEGA zones (Nordregio 2005) as well, that highlight the complexity of the European spatial structure and also the visualization of the core areas; they also highlight the increase in the differences between urban and rural areas and the differences between big cities and rural areas. Within the Nordregio (Nordregio 2005) project, the urban areas that have the potential to counterbalance the "European Pentagon" were analysed and 76 functional urban areas (FUAs Functional Urban Area) were classified into a four-level (actually five-level, by highlighting Paris and London) system including the metropolitan European growth districts (MEGA Metropolitan European Growth Area).

As already mentioned, two global cities were highlighted, namely London and Paris, that are global centres.

MEGA 1 group (17 city regions): Munich, Frankfurt, Madrid, Milan, Rome, Hamburg, Brussels, Copenhagen, Zurich, Amsterdam, Berlin, Barcelona, Stuttgart, Stockholm, Düsseldorf, Vienna and Cologne. Ten out of these cities are situated within the Pentagon area.

MEGA 2 group (8 city regions): Athens, Dublin, Geneva, Göteborg, Helsinki, Manchester, Oslo and Turin.

MEGA 3 group (26 city regions): Prague, Warsaw, Budapest, Bratislava (therefore four city regions can be found in the countries that joined the EU in 2004), Bern, Luxembourg, Lisbon, Lyon, Antwerp, Rotterdam, Aarhus, Malmö, Marseille, Nice, Bremen, Toulouse, Lille, Bergen, Edinburgh, Glasgow, Birmingham, Palma de Mallorca, Bologna, Bilbao, Valencia and Napoli.

MEGA 4 group (26 city regions, out of which 15 can be found in the new member states): Bucharest, Tallinn, Sofia, Ljubljana, Katowice, Vilnius, Krakow, Riga, Lódz, Poznan, Szczecin, Gdansk-Gdynia, Wroclaw, Timisoara, Valletta, Cork, Le Havre, Southampton, Turku, Bordeaux, Seville, Porto and Genoa (Nordregio 2005).

Besides these spatial structural descriptions and analyses, one cannot forget about the question of time as well, provided that the appreciation and depreciation of geographical areas happened not only in the past, but it goes on currently and probably it will continue in the future as well. This is supported by the shift of the economic priority southwards in the past decades, which modified the extension of the most popular form as well. Brunet also argues that the original shape of the "Blue Banana" extended southwards, therefore the form that existed in several interpretations ("North eastern arc", "German hump") even before, kept on changing, demonstrating the truth of Heraclitus stating that "change is the only constant".

In many cases, these are not the characteristics, nor the extension of the form describing spatial structure that are determinants, but the possibilities to link to the core areas and to the dynamic areas and the way and the kind of developments that makes it possible to utilize the advantages and the positive effects. Per capita income levels and economic growth rates are significantly higher in the countries that are situated close to the current centres of the world economy. Good market availability thus seems to be a great advantage for the employed in the globalized economies. Therefore an important question of the future development in the world economy refers to where the economic activity will be concentrated (Hospers 2003).

In the next sections we examine the background of the spatial structural relations and models described above more thoroughly with the use of four methods and with the help of spatial models, each representing a different approach to the problem.

Methods

Spatial moving average

The method of the spatial moving average can be used in the analysis of spatial phenomena and basic structure (Dusek 2001). In our analysis, our aim was to reveal stronger relationships with the help of moving averages. This can be done by finding the appropriate aggregation. In the case of a given elemental unit, the spatial moving average of the examined characteristic can be found by calculating the average of the values for the surrounding areas, defined based on the given topological characteristics in Equation. (1) (Haining 1978, Mur 1999)

$$M(x_i) = \frac{\Sigma (f_j \cdot x_j)}{\Sigma f_j} \tag{1}$$

for elements where $d(x_i; x_j) \leq m$

where $M(x_i)$ is the moving average of point i, $d(x_i, x_j)$ is the distance between the centres of i and j spatial units and m is the extension of the moving average (radius). x_j refers to the value to be averaged belonging to the jth observation, i.e., per capita GDP, and f_j is the frequency or weight belonging to the jth observation. If the moving average of per capita GDP is calculated, it is the population.

About gravity and potential models – Relationship between space and weight, separating potential

One of the methods most frequently applied to examine spatial structure in the literature is the potential model. The general formula for potential models is given in Equation (2) (see for example in Hansen 1959, Rich 1980, Isard 1999) :

$$A_i = \sum_j D_j \cdot F\left(c_{ij}\right)$$

$$(2)$$

where A_i is the potential of a spatial unit i, D_j is the mass of the spatial unit j, c_{ij} is the distance between the centre of i and j units (straight line distances) and $F(c_{ij})$ is the resistance factor (function).

The potential therefore is calculated from the sum of its own and internal potentials (Pooler 1987) using Equation (3) .

$$\sum A_i = SA_i + BA_i$$

$$(3)$$

where ΣA_i is the overall potential of the area i, SA_i is its own and BA_i is the internal potential. The potential value in a given point is therefore determined by the internal and own potential (the sum of its own mass and the effect of its own area size). The own potential refer to the effect of the spatial unit i on its own potential, while internal potential shows the impact of all other units on the potential of unit i.

Based on the topology of the geometry of potential models, one can conclude that whichever model is used, a common point is that they measure the effects of the position of a space range and the size distribution of the masses as described in Equation (4). The position of the space range is basically defined by the geographical position. This means that for a given potential value, it is not possible to decide whether it is a consequence of the position of the favourable/unfavourable (settlement, regional) structure, position or masses, of the area size or of the effect of its own mass. Therefore, we aim at separating these effects, describing the share of the parts in the overall potential values and introducing territorial differences.

$$\sum A_i = BA_i + SA_i = U_i^{mass\ distribution} + U_i^{location} + U_i^{mass\ weight} + U_i^{area\ size}$$

In an arbitrary point of the space, the effect of the potential derived from the spatial location refers to the value that could have been provided that the masses are the same in each of the specified territorial units, as in Equation (5).

$$U_i^{location} = \sum_j \frac{\left(\frac{\sum_{k=1}^{n} m_k}{n} \right)}{f(d_{ij})} \qquad (5)$$

where i, j, k are territorial area or units, m_k is "mass" in the kth territorial unit, which can be GDP, population etc.; n is the number of territorial units included in the analysis and $f(d_{ij})$ is the resistance factor, function.

The effect of mass distribution in an arbitrary point of the space is the value-difference between the internal potential and the location potential at the given point:

$$U_i^{mass\ distribution} = BA_i - U_i^{location}$$

(6)

The effects of area size (Equation (7)) and own mass (Equation (8)) can be interpreted accordingly in the case of their own potentials (the signs are the same as above).

$$U_i^{area\ size} = \frac{\left(\dfrac{\sum_{i=1}^{n} m_i}{n}\right)}{f(d_{ii})}$$

(7)

$$U_i^{own\ mass} = SA_i - U_i^{area\ size}$$

(8)

where m_i is "mass" in the ith territorial unit, which can be GDP, population etc.; n is the number of territorial units included in the analysis, d_{ii} is the distance within the spatial unit, which is calculated in a way that the area of a unit is considered to be circle. The radius of this circle is equal to the own distance. $f(d_{ii})$ is the resistance factor or function

Gravity models and examination of the spatial structure

After separating the potential models as described above, the other approach to examine spatial structure is about gravity models that are based on the application of forces. With the approach that we present here, one can assign attraction directions to the given territorial unit. This method complements and specifies the view of spatial structure described by the potential models.

The law of general mass attraction, Newton's law of gravitation (1686), states that any two point masses attract each other by a force that is proportional to the product of the two masses (these are heavy and not powerless masses) and is inversely proportional to the square of the distance between them (Budó 1970):

$$F = \gamma \cdot \frac{m_1 \cdot m_2}{r^2} \qquad\qquad (9)$$

where the proportionality measure γ is the gravitational constant (regardless of space and time).

If the radius vector from point mass 2 to point mass 1 is signed with r, then the unit vector from point 1 to point 2 is —r and therefore the gravitational force applied on point mass 1 due to point mass 2 is (MacDougal 2013):

$$\vec{F}_{1,2} = -\gamma \cdot \frac{m_1 \cdot m_2}{r^2} \cdot \frac{\vec{r}}{r} \qquad\qquad (10)$$

A gravitational force field is definite if the direction and the size of the field strength (K) can be defined at each point of the given field. To do so, provided that K is a vector, three pieces of data are necessary in each point (two in the case of a plain), such as the rectangular components K_x, K_y, K_z of the field strength as the function of the place. Many force fields, however, like the gravitational force field, can be described in a much simpler way, that is, instead of three, using just one scalar function, the so-called potential (Figure 6). (Budó, 1970)

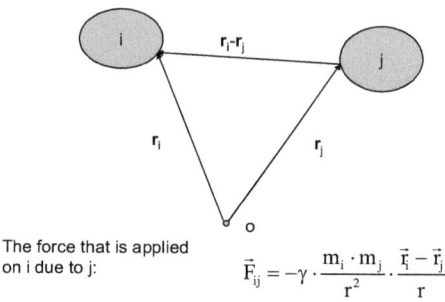

The force that is applied on i due to j:

$$\vec{F}_{ij} = -\gamma \cdot \frac{m_i \cdot m_j}{r^2} \cdot \frac{\vec{r}_i - \vec{r}_j}{r}$$

Figure 5: Calculation of the gravitational force (source: own compilation)

Potential is similarly related to field strength than force or potential force to strength. If in the gravitation field of K field strength, the trial mass, on which a force of $F=mK$ is applied, is moved to point B from point A by force $-F$ (without acceleration) along with some curve, then work of $L = -\int_A^B F_s ds$ has to be done against force F based on the definition of work. This work is independent of the curve from A to B.

Therefore it is the change of the potential energy of an arbitrary trial mass: $L = E_{potB} - E_{potA} = -\int\limits_A^B F_s ds = -m\int\limits_A^B K_s ds$.

By dividing by m, the potential difference between points B and A in the gravitational space is: $U_B - U_A = -\int\limits_A^B K_s ds$

By utilizing this relation, in most of the social scientific applications of the gravitational model the space primarily was intended to be described by only one scalar function (see for example the potential model) (Kincses–Tóth 2012), while in the gravitational law, it is mainly the vectors characterizing the space that have an important role. The main reason for this is that the arithmetic operations with numbers are easier to handle than calculations with vectors. In other words, for work with potentials, solving the problem also means avoiding calculation problems.

Even if potential models often show properly the concentration focus of the population or GDP and the space structure, they are not able to provide any information on the direction towards which the social attribute of the other regions attract a specified region and on the force with which they attract it.

Therefore, by using vectors we are trying to demonstrate in which direction spatial units are attracted by other units in the economic space compared to their real geographical position. With this analysis, it is possible to reveal the centres and fault lines representing the most important areas of attractiveness and it is possible to visualise the differences among the gravitational orientation of the spatial units.

In the traditional gravitational model (Stewart 1948) the "population force" between i and j are expressed in D_{ij}, where W_i and W_j are the populations of the units (settlements, regions, etc.), d_{ij} is the distance between i, and j and g is the empirical constant:

$$D_{ij} = g \cdot \left(\frac{W_i \cdot W_j}{d_{ij}^2}\right) \qquad (11)$$

Applying spatial potential is often not exactly the gravity law, but by analogous procedures is used, with help of different potential functions defined.

What of these we apply the

$$L = \sum_{i,j} g \cdot \left(\frac{W_i \cdot W_j}{d_{ij}^k}\right) \qquad (12)$$

k=1, 1,5, 2,.....

shaped in more detail. These potentials are converted into forces according to above detailed formula, which based on connection of the forces and potentials.

With the generalisation of the above formula, the following relationship is given in Equation (13) and (14):

$$D_{ij} = \left| \bar{D}_{ij} \right| = \frac{W_i \cdot W_j}{d_{ij}^c}$$

$$\bar{D}_{ij} = -\frac{W_i \cdot W_j}{d_{ij}^{c+1}} \cdot \bar{d}_{ij}$$

(13, 14)

where W_i and W_j indicate the masses taken into consideration, d_{ij} is the distance between them and c is the constant, which is the change in the intensity of the inter-territorial relations as a function of the distance. With the increase of the power, the intensity of the inter-territorial relations becomes more sensitive to the distance and at the same time, the importance of the masses gradually decreases (see Dusek 2003).

With this extension of the formula, not only the force between the two units but also its direction can be defined. In the calculations, it is worth dividing the vectors into x and y components, and then summarising them separately. In order to calculate this effect (the horizontal and vertical components of the forces), the necessary formulas can be deducted from Equation 15:

$$D_{ij}^X = \frac{W_i \cdot W_j}{d_{ij}^{c+1}} \cdot (x_i - x_j)$$

(15)

$$D_{ij}^Y = \frac{W_i \cdot W_j}{d_{ij}^{c+1}} \cdot (y_i - y)$$

(16)

where x_i, x_j, y_i, y_j are the centroids of spatial units i and j.

If, however, the calculation is carried out for each unit included in the analysis, the direction and the force of the effect on the given territorial unit can be defined using Equation (17) and (18).

$$D_{ij}^{X} = -\sum_{j=1}^{n} \frac{W_i \cdot W_j}{d_{ij}^{c+1}} \cdot (x_i - x)$$

$$D_{ij}^{Y} = -\sum_{j=1}^{n} \frac{W_i \cdot W_j}{d_{ij}^{c+1}} \cdot (y_i - y)$$

(17,18)

With these equations, in each territorial unit, the magnitude and the direction of the force due to the other units can be defined. The direction of the vector assigned to the units determines the attraction direction of the other units, while the magnitude of the vector is related to the magnitude of the force. In order to make visualisation possible, the forces are transformed to proportionate movements in Equation (19) and (20):

$$x_i^{mod} = x_i + \left(D_{ij}^{X} * \frac{x^{max}}{x^{min}} * k \frac{1}{\dfrac{D_{ij}^{X\,ma}}{D_{ij}^{X\,mi}}} \right)$$

(19)

$$y_i^{mod} = y_i + \left(D_{ij}^{Y} * \frac{y^{max}}{y^{min}} * k \frac{1}{\dfrac{D_{ij}^{Y\,ma}}{D_{ij}^{Y\,mi}}} \right)$$

(20)

where X_i mod and Y_i mod are the coordinates of the new points modified by gravitational force, x and y are the coordinates of the original point set, their extreme values are x_{max}, y_{max}, x_{min}, and y_{min}. $D_{ij}{}^X$ and $D_{ij}{}^Y$ are the forces along the axes, their extreme values are $D_{ij}{}^{Xmax}$, $D_{ij}{}^{Xmin}$, $D_{ij}{}^{Ymax}$, $D_{ij}{}^{Ymin}$ and k is a constant, in this case its value is 0.5. We obtained this value as a result of an iteration process.

It is possible a different approach of the linear projection, which follows we mention as the second method. The direction of the vector in this case is determined by the attraction of the other unit area, while the length of the vector will be line with magnitude of the force. For reasons of mapping and illustration, the received forces we transform into displacements according to the following manner (21-22 formula):

$$x_i^{mod} = x_i + \left(D_{ij}^X * (x^{max} - x^{min}) * \frac{1}{D_{ij}^{Xmax}} \right) \qquad (21)$$

$$y_i^{mod} = x_i + \left(D_{ij}^y * (y^{max} - y^{min}) * \frac{1}{D_{ij}^{ymax}} \right) \qquad (22)$$

X_i mod and Y_i mod are new coordinates, what the gravitational force modified, x and y are the original point set coordinates, these extreme values of x_{max}, y_{max}, a x_{min}, y_{min}, are the forces along the axis, D_{ij}^{xmax}, and D_{ij}^{ymax} are the maximum value of D_{ij}.

We want to avoid the possible type of impact effects with simultaneous applicability of the two projection methods, which is intended to guarantee the independence of the projection results.

Then it is worth comparing the new point set with the original one. This can naturally be done with visualisation, but in the case of such a large number of points, this alone probably does not provide a really promising result. Much more favourable results can be obtained by applying bidimensional regression analysis (see the equations related to the Euclidean version in Table 1) (about the bidimensional regression see Tobler 1994, Kare–Samal–Marx 2010, Nakaya 2010, Symington– Charlton–Brunsdon 2002).

Table 1: The equations of the bidimensional Euclidean regression

1. The regression equation	$\begin{pmatrix} A \\ B \end{pmatrix} = \begin{pmatrix} \alpha_1 \\ \alpha_2 \end{pmatrix} + \begin{pmatrix} \beta_1 & -\beta_2 \\ \beta_2 & \beta_1 \end{pmatrix} * \begin{pmatrix} X \\ Y \end{pmatrix}$
2. Scale difference	$\Phi = \sqrt{\beta_1^2 + \beta_2^2}$
3. Rotation	$\Theta = \tan^{-1}\left(\dfrac{\beta_2}{\beta_1}\right)$
4. β_1	$\beta_1 = \dfrac{\sum(a_i - \bar{a})*(x_i - \bar{x}) + \sum(b_i - \bar{b})*(y_i - \bar{y})}{\sum(x_i - \bar{x})^2 + \sum(y_i - \bar{y})^2}$
5. β_2	$\beta_2 = \dfrac{\sum(b_i - \bar{b})*(x_i - \bar{x}) - \sum(a_i - \bar{a})*(y_i - \bar{y})}{\sum(x_i - \bar{x})^2 + \sum(y_i - \bar{y})^2}$
6. Horizontal shift	$\alpha_1 = \bar{a} - \beta_1 * \bar{x} + \beta_2 * \bar{y}$
7. Vertical shift	$\alpha_2 = \bar{b} - \beta_2 * \bar{x} - \beta_1 * \bar{y}$
8. Correlation based on the error terms	$r = \sqrt{1 - \dfrac{\sum\left[(a'_i - a_i)^2 + (b_i - b'_i)^2\right]}{\sum\left[(a_i - \bar{a})^2 + (b_i - \bar{b})^2\right]}}$
9. The resolution difference of a square sum	$\sum\left[(a_i - \bar{a})^2 + (b_i - \bar{b})^2\right] =$ $\sum\left[(a'_i - \bar{a})^2 + (b'_i - \bar{b})^2\right] + \sum\left[(a_i - a'_i)^2 + (b_i - b'_i)^2\right]$ SST=SSR+SSE
10. A'	$A' = \alpha_1 + \beta_1(X) - \beta_2(Y)$
11. B'	$B' = \alpha_2 + \beta_2(X) + \beta_1(Y)$

Source: Tobler (1994) and Friedman–Kohler (2003) cited by Dusek 2012, 64.

Where x and y refers to the coordinates of the independent form, a and b sign the coordinates of the dependent form, a' and b' are the coordinates of the independent form in the dependent form. α_1 refers to the extent of the horizontal shift, while α_2 defines the extent of the vertical shift. β_1 and β_2 are used to determine the scale difference (Φ) and Θ is the rotation angle. SST is total sum of squares, SSR is sum of squares due to regression, SSE is explained sum of squares of errors/residuals that is not explained by the regression).

To visualise the bidimensional regression, the Darcy program can be useful (Vuidel 2009).The grid fitted to the coordinate system of the dependent form and its interpolated modified position make it possible to further generalise the information about the points of the regression.

The arrows show the direction of movement and the grid colour refers to the nature of the distortion.

Spatial autocorrelation analysis

The research of spatial autocorrelation, which often only called LISA (Local Indicators of Spatial Association) in the international scientific literature, started following the path breaking work of Luc Anselin (1995). Local auto-correlation indices were already used by several studies in Hungary (Tóth 2003, Bálint 2011, Tóth–Kincses 2012). With the introduction of Moran's I, Luc Anselin (1995) developed the Local Moran's I statistic, which is one of the most commonly used methods to quantify and visualize spatial autocorrelation; in our article we used it to explore the spatial economic relations of large cities. Using the designation (1996) of Getis and Ord, I is defined as (Formula 23):

$$I_i = \frac{(Z_i - \overline{Z})}{S_z^2} \cdot \sum_{j=1}^{N} \left[W_{ij} \cdot (Z_i - \overline{z}) \right].$$

(23)

where Z is the average of all units, Z_i is the value of unit I, S_z^2 is the dispersion of variable z for all observed units and

W_{ij} is the distance weighting factor between i and j units, which comes from the W_{ij} neighborhood matrix (basically $W_{ij} = 1$ if i and j are neighbors and 0 if they are not).

If we receive the Local Moran's I value, the negative values mean a negative autocorrelation and the positive ones a positive autocorrelation. At the same time, the function has a wider range of values than the interval of -1; +1. The indicator also has a standardized version, but now we do not deal with this. The Local Moran statistics is suitable to show the areas that are similar to or different from their neighbours. The bigger the Local Moran I value, the closer the spatial similarity. However, in case of negative values, we may conclude that the spatial distribution of the variables is close to a random distribution. During our work, it's worth to compare the results of the Local Moran statistic with the initial data in order to be able to examine whether the high degree of similarity is caused by the concentration of the high or low values of the variable (Moran Scatterplots). As a first step, on the horizontal axis of a graph the standardized values of the observation units were plotted, while on the y-axis the corresponding standardized Local Moran's I values (average neighbour values) were plotted. The scatterplot puts the municipalities into four groups according to their location in the particular quarters of the plane:

1. High-high: area units with a high value, where the neighbourhood also has a high value.

2. High-low: area units with high value, where the neighbourhood has a low value.

3. Low-low: area units with low value, where the neighbourhood also has a low value.

4. Low-high: area units with low value in which the neighbourhood has a high value.

The odd-numbered groups show a positive autocorrelation, while the even-numbered groups a negative one.

Of the local spatial autocorrelation indices, it is really appropriate to choose a Local Moran I if you search for spatially outlying values. Namely, on the one hand, it shows where the high / low values are grouped in the space (HH–LL) and, on the other hand, it shows where those territorial units are, which are significantly different from their neighbours (HL–LH).

Results

Spatial moving average

In this case, the level of aggregation is defined in a way to ensure its link to a territorial level that has currently been analysed. This was the NUTS1 level in our analysis. This territorial level was measured at its average extension, since supposing that the average area of the NUTS1 regions is a circle, a circle with 70 km radius is given. We carried out the calculations applying a 70 km radius, but we still judged our result to provide too fragmented picture. We presumed that the reason for this can be the relatively large dispersion among the areas of the NUTS1 level regions. Therefore we considered it more appropriate to define the radius of the moving average as 100 km; then, by increasing it by 20 km, we carried out the calculations up to a radius of 200 km. The reason for increasing the radius is that the higher the degree of aggregation, the higher the abstraction is, although after a certain size the loss of information increases as well.

The resulting map is much less fragmented compared to the base data, thus providing a possibility to carry out a more detailed analysis. Based on the map (Figure 6), we can conclude that the regions in the most favourable position in Europe – the engines of the economy – emerge from the examined areas like islands. These regions are primarily certain southern provinces in Germany, the regions of Rome and Northern Italy; the Northern part of Switzerland, a considerable part of Austria, the agglomerations of London and Paris, most of the area of the Benelux countries and of Denmark, the core area including a considerable number of the regions of each Scandinavian country. Besides these, outstanding values can only be found in the case of some regions. Such outstanding islands can be South Ireland (O'Reilly 2004), North Spain (Basque Country) and South Scotland. Considering Eastern European regions, the effect of the "Iron Curtain" is still determinant. In this part, these are mainly the agglomerations of the capitals (especially Bratislava) that emerge from their surrounding; the degree to which they lag behind the above mentioned regions is, however, considerable. Out of the regions of the countries belonging to the formerly socialist block, only a few have the potential to link to the mentioned core areas. In this context, only some regions of Slovenia (especially Ljubljana (see

Ravbar, Bole, and Nared 2005)) and the Czech Republic can be highlighted as positive examples.

With the above-described increase of the radius, we intended to increase the degree of abstraction. We increased the radius by 20 km each time, which made the results smoother. The outstanding areas are isolated from their surroundings; therefore, the main centres kept crystallizing. The results of the 200 km moving average can be seen on Figure 7.

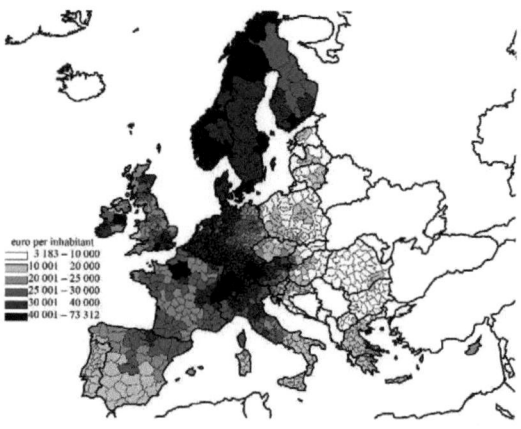

Figure 6: Spatial moving average of per capita GDP (2011) calculated with 100 km radius. (source: own compilation)

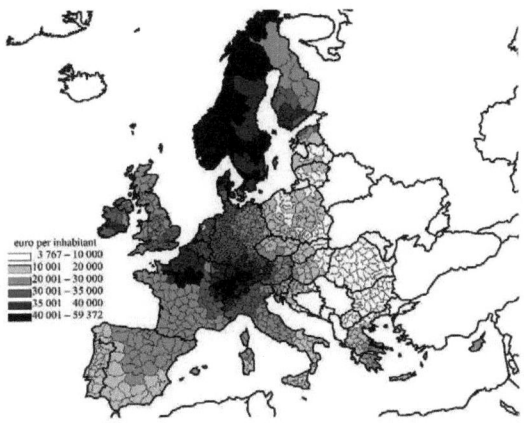

Figure 7: Spatial moving average of per capita GDP (2011) calculated with 200 km radius. (source: own compilation)

In case of the changes it can be seen (Figure 8 and 9.) that the most significant increase is visible at the states that joined in 2004 to the European Union, within these stand out in Romania and the Baltic regions. In the western part of Europe some Spanish and Swedish region emerging, but their lags behind the previously mentioned regions. The regions showing the most positive change were supported by the EU Structural Funds displacement, which rate was not enough to turn to the previous conditions of the spatial structure significantly.

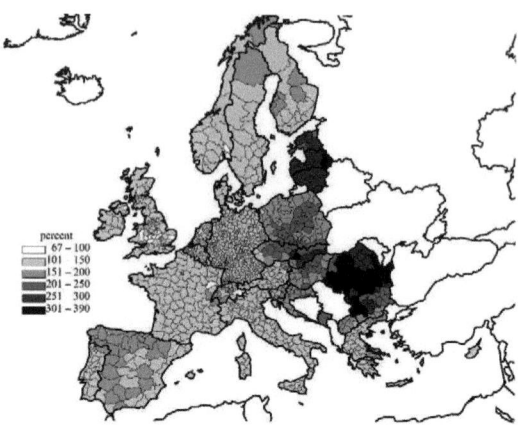

Figure 8: Changes in the spatial moving average of per capita GDP (2011/2001) calculated with 100 km radius. (source: own compilation)

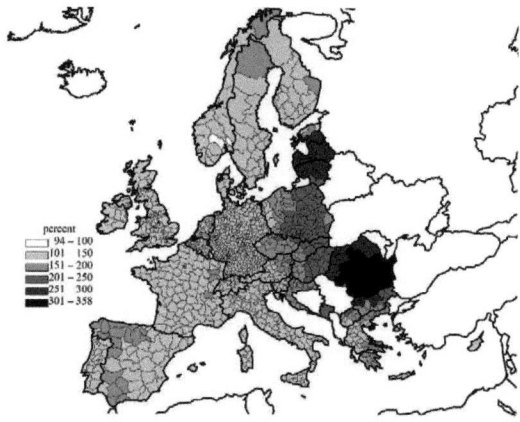

Figure 9: Changes in the spatial moving average of per capita GDP (2011/2001) calculated with 200 km radius. (source: own compilation)

Results of potential analysis

According to our potential analysis, the region in the most favourable position (in regard to the overall potential) within the European Union is Paris, followed by Inner London and Hauts-de-Seine (Figure 10). In general, it can be concluded that regions in the most favourable positions are the central regions of France and the regions of South England, the Netherlands, Belgium, Switzerland, and North Italy, and West Germany. The potential decreases gradually from the indicated core areas towards the peripheries. Our results justify the Blue Banana spatial structural model (Brunet 1989) and its extension to a certain extent (Kunzmann 1992).

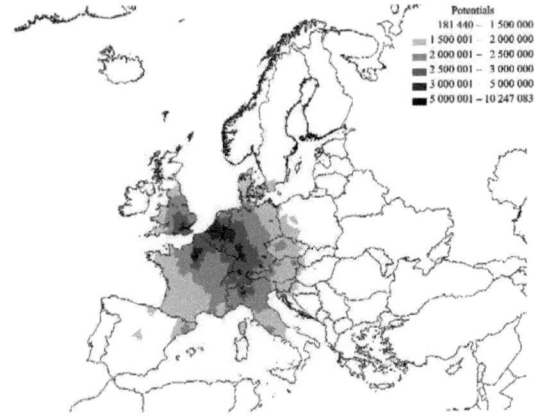

Figure 10: The potential values of regional GDP, 2008.
(source: own compilation)

Let us review the effects of the potential components. Within the potential, the effect of spatial location reflects the core-periphery relations; that is, the effect keeps decreasing as we move away from the geographical centre (Figure 11). The effect of the position is positive in each case, meaning that it always contributes to the overall potential. The effect of spatial location is the most important component within the overall potential for each of the regions. This means that the basic spatial structural relations – demonstrable with the help of the potential model – are determined mostly by the core-periphery relations in Europe; and other, later described components are able to modify this basic structure only slightly. Out of the known spatial structural models, this form is most similar to the European Pentagon (ESDP 1999) (see Figure 11).

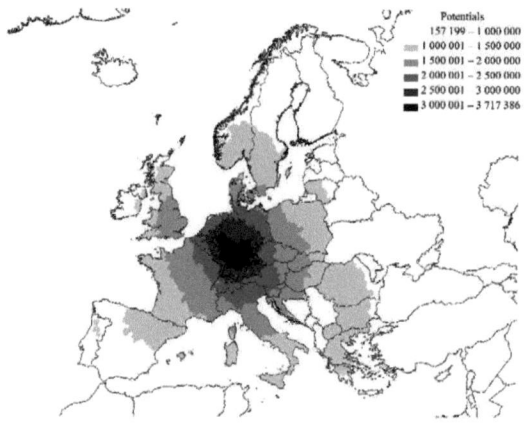

Figure 11: The role of spatial location in the potential values of the regional GDP. (source: own compilation)

As for the mass distribution, the catchment areas of London and Paris are outstanding (Figure 12). The effect of mass distribution contributes to the overall potential, contrary to the previous component, both negatively and positively. Out of the 1,378 examined regions, in 833 cases the sign is negative, while it is positive in the remaining 545 cases.

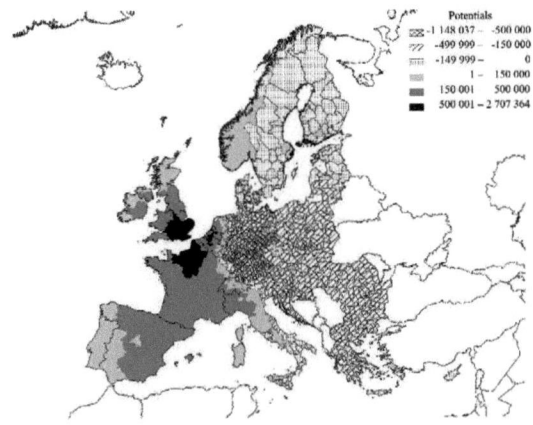

Figure 12: The role of mass distribution in the potential value of regional GDP. (source: own compilation)

The next two components (area size and the own mass of the given region) constitute the own potential part of the potential model. In the first case, we deal with the area size (Figure 13). Provided that the area of the given region is taken into consideration when calculating own potential (when we calculated own distance), the value of this component changes to the extent of the areas of the regions. The sign of the area size is always positive and its extent is inversely related to the area of the region. Thought we did not use population data, we can conclude that the value of this component refers primarily to urbanisation, since the regions with smaller area are big cities in most of the cases.

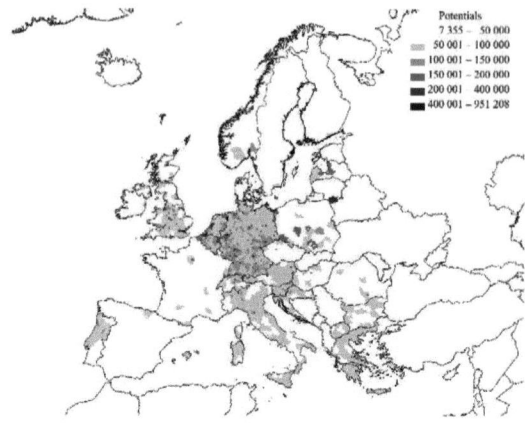

Figure 13: The role of area size in the potential value of regional GDP. (source: own compilation)

Finally, the last component is the own mass of the given region (Figure 14). Its sign can also be either negative or positive.

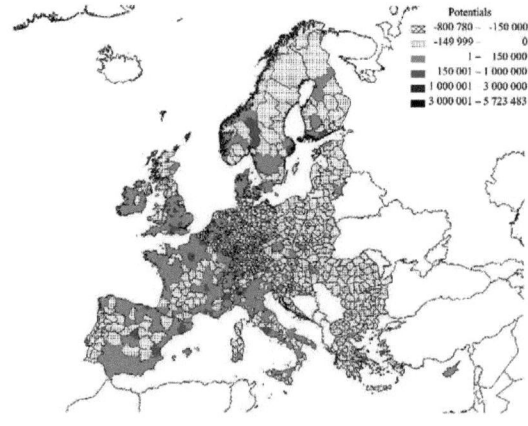

Figure 14: The role of own mass in the potential values of regional GDP. (source: own compilation)

In total, we can conclude that the different spatial structural models available in the literature can be synthetised by dividing the potential models into parts. The division into axes and zones can be shown in the analyses of spatial position and mass distribution, while the polycentric view can be linked to area size and to own mass. They visualise the real space structure side by side, complementing each other. By dividing the potential models into parts, the above described spatial structural ideas that are present in the space at the same time can be standardised.

Result of bidimensional analysis

Results at NUTS 1 level

Our analysis can be carried out at the NUTS1, 2, and 3 levels. The comparison of the results with those of bidimensional regression can be found in Table 2.

Table 2: Bidimensional regression between gravitational and geographical spaces.

Level	r	α1	α2	β1	β2
NUTS1	0.91	0.19	0.69	0.99	0.00
NUTS2	0.97	0.04	0.15	1.00	0.00
NUTS3	0.99	0.13	-0.04	1.00	0.00

Level	Φ	Θ	SST	SSR	SSE
NUTS1	0.99	0.00	20 430	19 849	582
NUTS2	1.00	0.00	54 121	53 484	638
NUTS3	1.00	0.17	139 884	139 847	37

Source: own calculation.

Figure 15: Directions of the distortion of gravitational space compared to geographical space for the European regions (NUTS1) (source: own compilation)

As the results show, the lower the level that is used for the analysis, the smaller the deviation of the gravitational point form is from the original structure. This is proven by the correlation and by the sum of squared deviations and their components. Because of the mass differences among the regions, the analysis carried out at different territorial levels shows results that are different in their nature even if they are similar in many aspects of their basic structure. That is why we decided to carry out the analysis at each territorial level in order to examine the different levels of the spatial structure. We visualised our results, and we drew the following conclusions. The analysis carried out at the NUTS 1 level contains only the most general relations (Figure 15). These general relations, however, are not sufficient to carry out a deeper analysis of the spatial structure. That is why it is necessary to go on to the NUTS 2 level. In this case, as shown in Figure 16, regional concentrations can unambiguously be seen, and we consider these to be the core regions. When the analysis is carried out at the NUTS3 level, a "lower" level of spatiality can be modelled. In this case, not the most important macro level structures (the basic core periphery relations), but the mezzo level elements, the deeper spatial relations can be revealed. It is not possible to publish here the map indicating the deeper relations of the regions; here we only summarise the most important conclusions regarding these deeper relations.

Comparison of the gravity methods at NUTS 2 level

Our analysis is carried out at the NUTS 2 level. The comparison of the results (between real and modified coordinates) with those of bidimensional regression can be found in Table 3.

Table 3: Bidimensional regression between gravitational and geographical spaces.

Methods	r	α_1	α_2	β_1	β_2
1st method	0.96	0.01	0.04	1.00	0.00
2nd method	0.94	0.07	0.30	0.99	0.00

Methods	Φ	Θ	SST	SSR	SSE
1st method	1.00	0.00	35 223	34 856	367
2nd method	0.99	0.00	55 829	53 141	2 687

Source: own calculation.

As shown in Table 3, the difference between the two methods is not significant and with some restrictions, the results of visualization are considered independent of projection. In case of the first method, the relationship is closer between gravity and the geographical coordinates. The reason is that in first case is smaller the value of the horizontal and vertical displacement and the range difference and the rotation angle too. As a result, of course, the differences of sum of squares are significantly smaller.

Figure 16: Directions of the distortion of gravitational space compared to geographical space for the European regions (NUTS 2), 1st method (source: own compilation)

As shown in Figure 16, regional concentrations can be seen unambiguously, and we consider them to be the core regions. Based on the analysis carried out at the NUTS 2 level, five gravitational centres, slightly related to each other, can be found in the European space. Gravitational centres are the regions that attract other regions and the gravitational movement is toward them. These three centres or cores are: 1) the region including Switzerland, Northern Italy and the French regions neighbouring Switzerland; 2) the region including the Benelux countries, Paris and it's surrounding and most of the regions of England; 3) the region including Berlin and the Brandenburg; 4) the region including Central Italy and 5) the region including Languedoc-Roussillon, Midi-Pyrénées and Catalonia. Mainly these core areas have an effect on the regions of the examined area.

We find that the key element of the economic spatial structure of Europe is the structure reflected by the Blue Banana and the German Hump theory.

The issue of distance

We presented a variety of distance approach in our models. This is, of course, a common practice in social science studies, even if the original physical analogy is slightly different, because there the number 2 is applied as exponent, what is the law of physics (the value is 2, not 1.99 or 2.01). Our models are not exactly gravitational models, but based on gravitational analogy. That's why we calculated the exponents taking into account in order distance-dependence to investigate the role of mass and the distance on the European gravitational field.

As Tamás Dusek (2003) remarked in the work of gravitational models: "The intensity of the relationship among the area will be distance-sensitive, as the distance power increases, parallel to the masses gradually decreasing importance"

Table 4: The correlation coefficients in the two methods,

taking into account different exponents

c	1st method	2nd method
0.0	0.752964844	0.693314218
0.5	0.738790230	0.922820959
1.0	0.859542280	0.790773055
1.5	0.860785077	0.725618881
2.0	0.860891879	0.717864602
2.5	0.860918153	0.715549296
3.0	0.860926003	0.714371559

Source: own calculation.

The c values come from 13-14 formula. There are difficult to reconcile the constants values of $c = 0$, $c = 0.5$ and $c = 1$ with the traditional conceptions of spatial imaginative, but in any case, we see that when c values increase, the forces of impact area is reduced. This entails the convergence of correlation coefficients.

Analysis with different indicators at NUTS2 level

The following tests were conducted with other variables as well. Our analysis can be carried out at the NUTS2 level. The comparison of the results (between real and modified coordinates) with those of bidimensional regression can be found in Table 5.

Table 5: Bidimensional regression between gravitational and geographical spaces.

Indicator	r	α1	α2	β1	β2
Population	0.97	0.02	0.08	1.00	0.00
Employment	0.97	0.02	0.06	1.00	0.00
GDP	0.97	0.06	0.04	1.00	0.00

Φ	Θ	SST	SSR	SSE
1.00	-0.34	53 272	52 643	629
1.00	-0.75	51 959	51 446	493
1.00	0.62	51 974	51 480	494

Source: own calculation.

There is no significant difference in the gravitational shifts created using the different variables, which is indicated by the equally high values of the two-dimensional correlation (r). Its highest value can be one, which is reached when the exact coordinates of the points coincide with each other as a result of motion, rotation and rescaling. The minimal value of correlation is zero, which means that each point of a point pattern has the same coordinate. In our case, the difference between the geographical and gravitational coordinates is minimal. α_1 refers to the extent of the horizontal shift, while α_2 defines the extent of the vertical shift. The horizontal shift is the highest in the case of GDP, while the vertical shift is the highest in the case of the calculation using the population. β_1 and β_2 are used to determine the scale difference (Φ) and the rotation angle (Θ). In our analysis, a difference could only be found in the rotation angle. If Θ = 0, the XY coordinate system does not need to be rotated. If it is equal to zero, this means a clockwise rotation. It is also the case in our analysis that rotation is a bit higher for GDP than for the two other variables. Theoretically, decomposition of the total sum of squares is carried out in the same way as for a univariate case and also the notations are the same (SST: total sum of squares, SSR: sum of squares due to regression, SSE: explained sum of squares of errors/residuals).

Let' s see the results of bidimensional analysis! The arrows in Figure 24-27 show the direction of movement and the grid colour refers to the nature of the distortion.

The visualised analysis of the bidimensional analysis with three variables has slightly different results. Analysis using the population clearly highlights the most important actors of the European demographic space structure and the most populated, decisively urban areas (Figure 17). As for the number of the employed, the spatial picture is quite similar (Figure 18). Deviations in this aspect are slightly smaller and the extent of concentration is slightly more modest. As a result of the calculation using GDP the number of nodes decreases significantly (Figure 19). In the map with contour lines, the regions related to the so-called Blue Banana space structure – the formerly described economic engine of the European Union – emerge unambiguously. Within this area, two centres can be identified. On the one hand, the regions of South England, the Benelux states and North France make up the most important node, while in the case of the regions of North Italy and South Germany (and the related regions of Switzerland), a central position exists, but to a lower extent. This area emerges also as a result of the calculations carried out with the two other variables. In those cases, other areas are linked to it.

Figure 17: Directions of skewness of the gravitational space compared to the geographic space in the case of European (NUTS2) regions (mass factor: population) (source: own compilation)

Figure 18: Directions of skewness of the gravitational space compared to the geographic space in the case of European (NUTS2) regions (mass factor: number of the employed) (source: own compilation)

Figure 19: Directions of skewness of the gravitational space compared to the geographic space in the case of European (NUTS2) regions (mass factor: GDP) (source: own compilation)

Results at NUTS 3 level

One of the most important characteristics of the spatial structures that can be seen as a result of the NUTS 3 level analysis is that the importance of state borders is rather high, even though the method does not take borders into account (Figure 20).

Figure 20: Directions of the distortion of gravitational space compared to geographical space for the European regions (NUTS3) (source: own compilation)

Figure 21: Directions of the distortion of gravitational space compared to geographical space for the European regions (NUTS3), (British Isles). (source: own compilation)

Figure 22: Directions of the distortion of gravitational space compared to geographical space for the European regions (NUTS3), (Northern Europe) (source: own compilation).

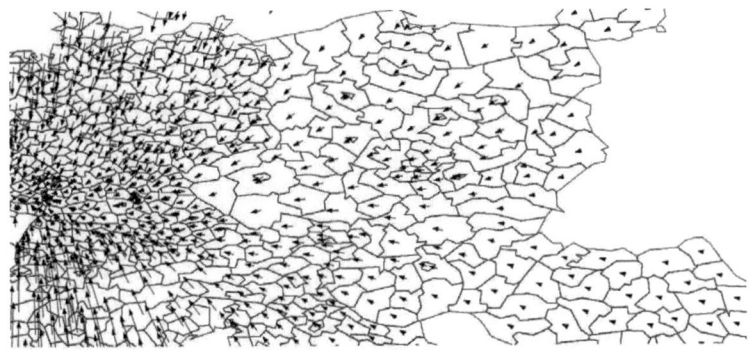

Figure 23: Directions of the distortion of gravitational space compared to geographical space for the European regions (NUTS3), (Eastern Central Europe) (source: own compilation).

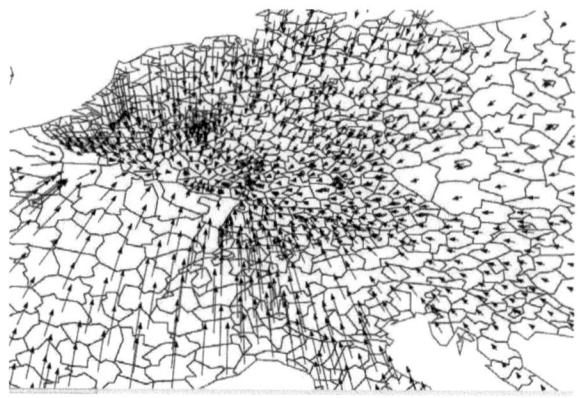

Figure 24: Directions of the distortion of gravitational space compared to geographical space for the European regions (NUTS3), IV. (Western Central Europe) (source: own compilation).

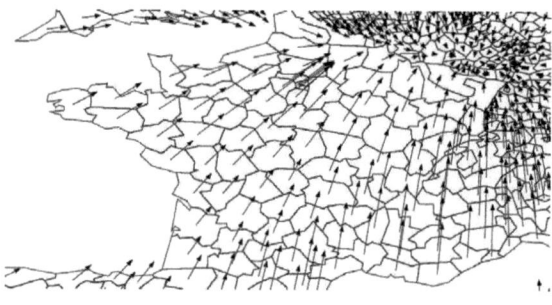

Figure 25: Directions of the distortion of gravitational space compared to geographical space for the European regions (NUTS3), (France) (source: own compilation).

Figure 26: Directions of the distortion of gravitational space compared to geographical space for the European regions (NUTS3), (Spain, Portugal) (source: own compilation).

Figure 27: Directions of the distortion of gravitational space compared to geographical space for the European regions (NUTS3), (Italy, Croatia, Switzerland) (source: own compilation).

Figure 28: Directions of the distortion of gravitational space compared to geographical space for the European regions (NUTS3), (Hungary, Greece, Croatia, Romania, Bulgaria, Macedonia. Cyprus) (source: own compilation).

The main fault lines of the gravitational space, where the gravitational shifts of the neighbouring regions are of opposite directions, were identified as:

1. seas, as important structural fault lines (the Adriatic and Tyrrhenian Sea and La Manche Channel)

2. the axis dividing Sweden into two parts

3. cities that emerging from their environment break the basic spatial directions (Madrid, Barcelona)

4. the axis dividing France into two parts.

The most important centres of gravitational space, where the gravitational movement of the neighbouring regions converges to the same direction, are:

1. the French-Swiss and German border area

2. the Dutch-German border area

3. the Danish-Swedish border area..

Analysis of the whole of Europe

In the analysis of the entire of Europe the population was only available indicator. The population of the European regions rose by nearly 4% from 761 million in 1990 to nearly 790 million in 2012. If we analyze the results of our population change related gravitational model (reference period: 1990-2012) then the most important gravitational nodes are in the southern part of Sweden and in Italy (in the area of the NORD-EST mega-region) (Figure 29). Further gravitational centres may be identified in the successor states of Yugoslavia, Albania, Pskov Oblast and in one part of the Spanish and Turkish regions. The regions affected by population decline were marked red. The significant part of Central and Eastern Europe was coloured red.

Figure 29: Changes in the gravity field based on population
figures in Europe, 1990–2012 (source: own compilation).

Result of Local Moran I

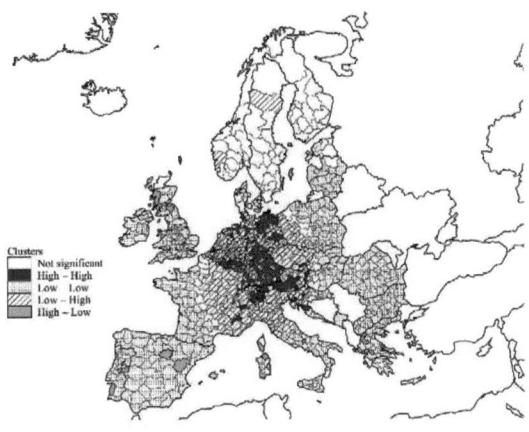

Figure 30: Figure of Local Moran I (source: own

compilation)

Due to the modifiable areal unit problem (Openshaw 1983), it was important that the clusters demarcations not only the level of development, that is, per capita income, note, but the ones that can be observed in GDP in that the per capita amount of population size regions. Thus we can handle different sized regional differences, and we can show within the European spatial structure of the most developed zones on it, which includes the High-High cluster. The calculations for this were performed using the software GeoDa with help of LISA with method rates.

The results in many respects identical with the previously noted, but it can also differ slightly. This model reflect the results of the Sunbelt zone, the German hump and the Blue Banana, The difference that both Ile de France and the Centre and Upper Normandy and Pays de la Loire is also ranked among the most favored regions. These regions only on the European Bunch of grapes model considered centrally located, but this study do not analyze more elements of this model

Analysis of the changing spatial structure

In the following section, we try to take into account the change of the structure. To do so, the gravity calculations are performed for 2000 and 2011. In order to measure changes, we compare and analyse the two gravity sets of points (2000 and 2011). The two-dimensional regression calculations are shown in Tables 6 and 7.

Despite the fact that the spatial structural changes take place over a decade or longer time, it was not possible to take into account the greater time interval. The reason for this was basically the recent changes of NUTS system, which makes it a suitable data, were available only between 2000 and 2011.

Table 6: Bidimensional regression between gravitational and geographical spaces

Years	r	α_1	α_2	β_1	β_2
2000	0.96	0.01	0.03	1.00	0.00
2011	0.96	0.01	0.04	1.00	0.00

Years	Φ	Θ	SST	SSR	SSE
2000	1.00	0.00	35 243	34 876	367
2011	1.00	0.00	35 223	34 856	367

Source: own calculation.

Table 7: Bidimensional regression between gravitational spaces

Years	r	α_1	α_2	β_1	β_2
2011/2000	1.00	0.00	0.01	1.00	0.00

Years	Φ	Θ	SST	SSR	SSE
2011/2000	1.00	0.00	35 223	35 223	0

Source: own calculation.

Our results show that there is a strong relationship between the two point systems; the transformed version from the original point pile can be obtained without using rotation (Θ = 0). No essential ratio difference between the two shapes is observed.

As we can see, in the examined time interval has not been a marked change in the spatial structure of the Europe. Despite this, it is worth examining the change in detail between 2000 and 2011, because during this period the changes may crystallize out the essential elements of the spatial structure modification. In term of change from 2000 to 2011, 7 gravity centres are shown on the map, indicated by shaded ellipses (Figure 31). They show a crucial part of the economic potential of big cities. Such hubs are the surroundings of Rome, Marseille-Zurich, Madrid, Toulouse, Brussels, Göteborg, Praha-Chemnitz etc. A gravity 'breakline' can be seen in Germany around Berlin and in central France.

In general, the change was not fundamental in the examined period but rather focused on only a few areas. These areas are parts of the Bunch of Grapes fields, which may show the increasing importance of this theory. However, there are fewer nodes or 'grapes' than the model predicts.

As far as the analysis of change is considered, the closest connection is to the Red Octopus model, because most gravity nodes were directly affected by the octopus arms. The analysis confirms the favourable position of certain regions, e.g. the Sunbelt zone, the Blue Banana. Our results confirm only partly the existence of the Central European Boomerang (Gorzelak 2012) especially as the author pointed out that development of the Berlin and it's catchment area a branch of this formation can occur.

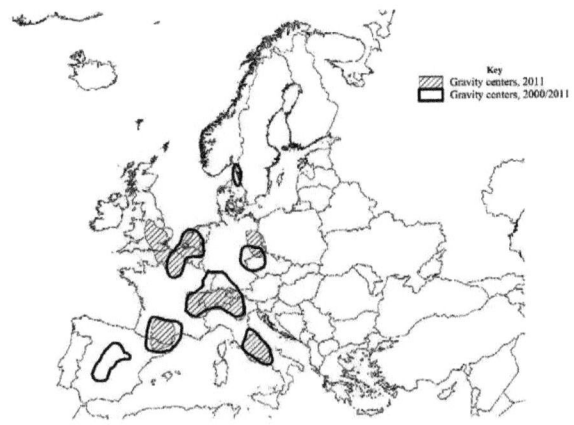

Figure 31: The results of the Gravity method at NUTS 2

level (source: own compilation)

Comparison of the applied methods, summary

The methods applied in this study used the same data and yielded different results. The comparison of the results methodologically is relatively difficult. Defining the core regions is easiest using the gravity analysis, provided that these are the regions that have converging spatial movements and that can be considered the main gravitational centres. These regions are shown in purple in Figure 32. In case of the moving average and the potential method the situation is a bit harder. In these cases, based on our data, the regions belonging to the upper quarter of the data series were considered core areas. In case of the Local Moran I the regions belonging to the High-High cluster were considered as core area. The visualised comparison based on this can be seen in Figure 32.

We can conclude that there are core regions based on each method that are not considered core regions on the basis of the other methods. In the case of the moving average, these are the Northern European regions, in the case of the potential method, it is Berlin, in the case of the gravitational method, these are the regions of Denmark and the southern part of Sweden, while in the case of Local Moran I these are the regions of East Germany. The intersection of the four models, however, can be seen, which definitely verifies the banana shape. The European core area, based on our analysis, still has the banana shape, like other authors concluded, but the different analyses highlight the existence of related regions that are moving to catch up. Furthermore, one of the most important results of our research is that the strongest determining element of the spatial structure is the spatial position component, obtained from the separation of the potential, which expresses the basic core–periphery relations. The other components can only slightly modify its effect; therefore the basic spatial relations can only be improved slightly by development tools.

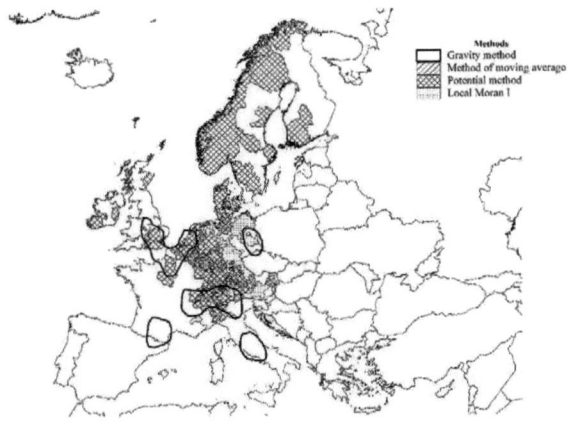

Figure 32: Comparison of the results of the four methods at NUTS 3 level. (source: own compilation)

Acknowledgement

„The described work was carried out as part of the TÁMOP-4.2.1.B-10/2/KONV-2010-0001 project in the framework of the New Hungarian Development Plan. The realization of this project is supported by the European Union, co-financed by the European Social Fund."

This work was supported by the János Bolyai Research Scholarship of the Hungarian Academy of Sciences.

References

Anselin, L. (1995): Local indicators of spatial association-LISA. *Geographical Analysis* 27 (2): 93–115.

Bálint, L. (2011): *A területi halandósági különbségek alakulása Magyarországon 1980–2006* KSH NKI, Budapest.

Brunet R. (1989): *Les villes europeénnes: Rapport pour la DATAR.* Reclus, Montpellier.

Budó Ágoston. (1970): *Kísérleti Fizika I.* Nemzeti Tankönykiadó, Budapest.

Dommergues P. (1992): The Strategies for International and Interregional Cooperation. *Ekistics* 352-353: 7–12.

Dusek Tamás (2001): A területi mozgóátlag. *Területi Statisztika* 41 (3): 215–229.

Dusek Tamás (2003): A gravitációs modell és a gravitációs törvény összehasonlítása. *Tér és Társadalom.* 17 (1): 41–57.

Dusek Tamás (2011): Bidimensional Regression in Spatial Analysis *Regional Statistics* 2 (1): 61–73.

ESDP (1999): *European Spatial Development Perspective.* Brussels. Europen Comission.(Adopted by the European Council of EU Ministers Responsible for Spatial Planning, in Potsdam, 10-11/05/99.

Faludi A. (2005): Polycentric territorial cohesion policy. In: Faludi A. (szerk.) Territorial Cohesion: An Unidentified political objective (Special Issue). *Town Planning Review* 76 (1) 107–118.

Friedman, A. – Kohler, B. (2003): Bidimensional Regression: Assessing the Configural Similarity and Accuracy of Cognitive Maps and Other Two-Dimensional Data Sets. *Psychological Methods*, 8(4): 468–491.

Getis, A. – Ord, J. K. (1996): Local spatial statistics: an overview. In: Paul Longley – Michael Batty: Spatial Analysis: Modelling in a GIS Environment. pp. 261–277. GeoInformation International: Cambridge, England.

Gorzelak G. (2012): *The Regional Dimension of Transformation in Central-Europe* Taylor & Francis, London.

Haining, R. P. (1978): The Moving Average Model for Spatial Interaction. *Transactions of the Institute of British Geographers* 3 (2): 202–225.

Hansen, W.G. (1959): How Accessibility Shapes Land-Use. *Journal of the American Institute of Planners.* 25 (2):. 73–76.

Hospers, G-J. (2003): Beyond the Blue Banana? Structural Change in Europe's Geo-Economy. *Intereconomics* 38 (2): 76–85.

Isard, W. (1999): Regional Science: Parallels from Physics and Chemistry. *Papers in Regional Science* 78 (1): 5–20.

Kare,S. – Samal,A. – Marx, D. (2010): Using bidimensional regression to assess face similarity.. *Machine Vision and Applications* 21 (3): 261–274.

Kilper, H. (ed.) (2009): *New Disparities in the Spatial Development of Europe* German Annual of Spatial Research and Policy. Springer, Berlin-Heidelberg.

Kincses Áron – Tóth Géza (2012): Geometry of Potential Models. *Regional Statistics* 2 (1): 74–89.

Kozma Gábor (2003): Térszerkezeti modellek Európában. In: Süli-Zakar István (szerk): *Társadalomföldrajz-Területfejlesztés*. pp 427-439., Debreceni Egyetem Kossuth Egyetemi Kiadó, Debrecen.

Kunzmann, K. R. (1992): *Zur Entwicklung der Stadtsysteme in Europa. Mitteilungen der Österreichischen Gesellschaft*. Wien.

Kunzmann, K.R. - Wegener, M. (1991): The pattern of urbanization in Europe. *Ekistics* 58: 282–291.

Kunzmann, K.R. (1996): Euro-megalopolis or Themepark Europe? Scenarios for European spatial development. *International Planning Studies* 1 (2): 143–163.

MacDougal, D. W (2013): *Newton's Gravity* Springer, New York | Heidelberg.

Meer L. van der (1998): Red octopus. In BLAAS W. (ed.) *A new perspective for European spatial development policies*. op. cit., pp. 9-19.

Mur, J. (1999): Testing for spatial autocorrelation: moving average versus autoregresive processes. *Environment and Planning A* 31(8):1371 – 1382.

Nakaya, T. (2010): Statistical Inferences in Bidimensional Regression Models. *Geographical Analysis* 29 (2): 169–186.

Nordregio (2005): *ESPON 1.1.1 Potentials for polycentric development in Europe* Nordregio/ESPON Monitoring Committee, Stockholm/Luxembourg http://www.espon.eu. [accessed February 2013].

Openshaw, S. (1983): *The modifiable areal unit problem* Geo Books, Norwick.

O'Reilly, G. (2004): Economic Globalisations: Ireland in the EU – 1973–2003 *Acta geographica Slovenica* 44 (1): 47–88.

Pooler, J. (1987): Measuring Geographical Accessibility: a Review of Current Approaches and Problems in the Use of Population Potentials *Geoforum* 18 (3): 269–289.

Ravbar, M. – Bole, D. – Nared, J. (2005): A creative milieu and the role of geography in studying the competitiveness of cities: the case of Ljubljana *Acta geographica Slovenica* 45 (2): 255–276.

Rich, D. C. (1980): *Potential Models in Human Geography*. Concepts and Techniques in Modern Geography no. 26 http://www.qmrg.org.uk/files/2008/12/26-potential-models-in-geography.pdf [accessed February 2013]

Schätzl L. (ed) (1993): *Wirtschaftsgeographie der Europäischen Gemeinschaft*. Uni-TB, Stuttgart.

Stewart, J.Q. (1948): Demographic Gravitation: Evidence and Application. *Sociometry* 11 (1-2):31–58.

Symington, A. – Charlton, M.E. – Brunsdon, C.F. (2002): Using bidimensional regression to explore map lineage. *Computers, Environment and Urban Systems* 26 (2): 201–218.

Szabó P. (2009): Európa térszerkezete különböző szemléletek tükrében. *Földrajzi Közlemények* 133 (2): 121–134.

Tobler, W. R. (1994): Bidimensional Regression. *Geographical Analysis* 26 (3): 187–212.

Tóth, G. (2003) Területi autokorrelációs vizsgálat a Local Moran I módszerével *Tér és Társadalom* 17 (4): 39–49.

Vandermotten C. – Halbert L. – Roelands M. – Cornut P. (2008): European planning and the Polycentric Consensus: Wishful Thinking? *Regional Studies* 42 (8): 1205–1217.

Vuidel, G. (2009): Darcy 2.0: module de comparaison spatiale *Spatial Simulation for the Social Sciences* http://spatial-modelling.info/Darcy-2-module-de-comparaison [accessed February 2013].

Wegener M. - Kunzmann K.R. (1996): New Spatial Patterns of European Urbanisation. In: Pumain D. - Saint-Julien T.(szerk.): *Urban Networks in Europe.* pp. 7-17. John Libbey, Paris.